The Grace I gave before the Grace I have

BY

Jason E. Parry

Copyright © 2024 by Jason Parry

Published by Hemingway Publishers
Cover design by Hemingway Publishers
ISBN: Printed in United States

The Grace I gave before the Grace I have

Intro

I really want to tell my story the way that makes sense to me.. so that's a lot of bouncing back and forth because that's how the Lord showed it to me. Most times, you don't really see why God is doing something or allowing it to happen until years later when it brings you strength and Him glory!

I imagine starting with my birth is beneficial to the story of my life and the many times the enemy tried to snuff it out.

I was born in Warren, Ohio.. on September 25,th.1976.. to Carolyn Sue Parry (at that time) and Jack Nelson Parry. I was my mother's 6th and final child.

Her only 4th one alive by the time that day came.

My sister Christine, who would have been my mom's oldest, hung on her umbilical cord and died at birth. My sister Kelly Joe, who was my mom's fourth child, died three days before her first birthday due to some children's disease called SIDS. The Dr said, "I'm not positive about that."

Seems the enemy was trying to take out all my mom's children.

I can tell you I was no exception. There were complications when she had me as well.

I was her biggest baby at 9 lbs 8 oz - 22 inches long with my face down.. so the back of my head wouldn't clear.

My mom told the Dr. if there was a chance of losing me.. he better take me and off to C-section we went..

I would later in life lose another sibling, and my mother's child survival rate before it was her time to go would become 50%..

But I'll talk more about that in a later chapter.

Chapter 1: The Enemy

I believe Satan tried to kill me several times..

I'll talk about each time in detail as the chapters go.. and where they fit in..

The first, I believe, was at birth, as I already mentioned. The next time wasn't much later..

As a five year old boy...I was wild and full of life, and energy was definitely not in short supply..

A friend of mine, Linda, and I had little childhood crushes on each other.. so we did things like pull hair and things of the sort.. not nice...but it was the 80s, lol.

We had a thorn bush next to our house on Philadelphia Street, where I grew up..

Linda pulled a branch from it one day, holding the thorny end for only God knows why..

I ripped it from her hands.. again...only God knows why..

A few weeks later, when her hand was healed..but her pride was still very bloody.. the enemy would use Linda to try his best to end me..

At five years old, with rage and pride in his corner. The enemy would speak through Linda's older sister Missy when she exclaimed...let's chase Jason with bricks.. and as they picked up red house bricks from their yard and chased me.. I ran.. only to be caught up with by Linda and have my skull smashed by a brick.. sending me to the emergency room to have 18 stitches put in my head..

My head healed, and my relationship with those girls stayed very much intact..

I hadnt yet seen how the Lord was going to move in my life, but forgiveness is going to bring Him glory, as it always has!

Chapter 2: Innocence gone

As I write this,, I actually struggle with titling this chapter..

This chapter will undoubtedly be one of the hardest, if not the hardest, chapters.

Philippians 4:13

New King James Version

13. I can do all things through Christ who strengthens me.

Fast forwarding to the innocent age of 10 years old...that traumatic thing that no parent wants for their child happened to me..

Mom, if you are reading this, I'm sorry I kept that from you all this time.

I didn't understand it until I was much older, and by then, it just seemed easier to keep silent..

I hadnt told anyone for years about that.

And to spare the details or any other unnecessary pain.. I will move on..

I will say it damaged me in many ways for many years.. especially spiritually..

The enemy landed a heavy blow to the man I would become..

It would take many, many years for me to understand Jesus when he said in

Matthew 11:28-30

New King James Version

28 Come to Me, all you who labor and are heavy laden, and I will give you rest. 29 Take My yoke upon you and learn from Me, for I am gentle and lowly in heart, and you will find rest for your souls. 30 For My yoke is easy, and My burden is light."

As a matter of fact, my life would start to become pretty dark in some ways.

One was a constant feeling that I was unwanted and, therefore, I should end my own life..

I never could get up the courage.. if you want to call it that... even to try..

And sometimes, I hated myself for that.. carrying around all that confusion and self-loathing..

Going through something like that leaves you confused about who you are and if you are good enough to even be alive..

That would ultimately set the pace for every relationship I had for the rest of my life until I met Jesus.. truly met Him..

You will know what I mean in future chapters.

Chapter 3: Slowly losing myself

As time went on, I became a people pleaser..

Anything to be accepted, loved, wanted, or at very least befriended.

I did this in my friendships, in relationships with girls, and even with the school bullies.

I was bullied a lot as a kid, starting about the 5th grade.. most of which I kept to myself like everything else.. it was a fear.. fear of defending myself (they smelled that all over me like wild dogs smell fear and prey) fear of telling on anyone cause they might get back at me..

I recall one incident when I was about 14 years old, and bullying didn't come from a student at school.. but an adult in the neighborhood.. a group of us kids was at my friend's Linda(yes, the one who split my melon with a brick) and Missy's house..

Well, they had a neighbor who didn't mow his lawn, and due to that, there was a rat infestation in and around their house.

On this particular day, he had happened to come outside right as we were talking about the rats.. and I told him he should cut his grass.. not realizing he was intoxicated.. he yelled some obscenities at me and went inside.. only to come quickly back outside with a gun in his hand..

It seemed to be a small revolver. It was a 38-special, to be my guess. I kinda hid behind the banister of Linda and Missy's back porch with my eyes peering just above the top rail so I could see what was going on and to keep an eye on the girls who had run behind the garage..

And as I looked over in his direction, I saw that he was screaming the entire time, and then all of a sudden, BANG! The gun went off, and my eyes were frozen on him.. then as fast as the gun went off.. he began to hop in place ...and with a quick turn and hoping limp type motion, he yelled at his wife, who was standing nearby.. GET THE DOOR.. GET THE DOOR.. And in that moment, I realized that, that guy just shot himself in the foot!

I laughed so hard at the time.. Looking back, tho.. I realized two things.. God saved my life that day.. and that man needed the Lord..

I never called the cops, and neither did my parents.. I told my mom, dad, and older brother what had happened. I was really raised with a... don't be a snitch background.. and a.. learn when to walk away and learn when to fight back.. Dad and older brother..

As I started to get older, I started to get over the physical bullying fear cause my bad attitude I was forming got me into more physical issues..

I had to learn to fight or get beat up way more than I was. Fighting started to interest me, and once I learned, I could win!

Unfortunately, all of that led me to lose myself and grow further away from who God desired me to be... Though I often think.. the pain was part of the plan.. He was letting me go through my trials and tribulations..

Romans 5:3-5

English Standard Version

3 Not only that, but we rejoice in our sufferings, knowing that suffering produces endurance, 4 and endurance produces character, and character produces hope, 5 and hope does not put us to shame because God's love has been poured into our hearts through the Holy Spirit who has been given to us.

See what I mean..? He didn't make these things happen necessarily.. we live in an evil fallen world.. But He didn't stop them from happening.. God is a gentleman, so He cannot and will not impose His will on any of our lives. But He does have a hope for us.

Jeremiah 29:11

New King James Version

11 For I know the thoughts that I think toward you, says the Lord, thoughts of peace and not of evil, to give you a future and a hope.

As I was losing myself and losing my hope, He was right there with me, giving up NO ground to the enemy!

Chapter 4: Frankie

I want to back up here a little bit to mention my best friend Frankie..

Now, truth be told, I could dedicate a whole book to Frankie and a few other neighborhood kids..

As a matter of fact, I will probably do so later on.

For now.. he gets a chapter..

I met Frankie in kindergarten.. I had just moved into the neighborhood and hadn't gotten out much before school started that year..

Frankie and I realized after just a few days of school that we actually lived on the same street.. in the same block.. I lived at the top of the hill on one side.. he was at the bottom on the other side of our street.. just before the railroad tracks..

We almost instantly became best friends..

Like brothers.. inseparable!!

We grew up together and carried each other's burdens.. No matter what! He had his pains, and I had mine, and sometimes we were the source of each other's pain.. and sometimes, we were what made the other one better..

We learned together about life and drinking.. drugs.. and girls.. and rock and roll.. and crashing motorcycles.. lol..

As I said.. I can go on and on about Frankie..

But sadly, we lost Frankie when he and I were just 23 years old..

A lady who had been drinking and driving had hit him as he was walking home from work one night.. killing him instantly..

The thing that always gets to me is that I don't know if he gave his life to Jesus.

We talked about it many times when we were younger, but I don't think back to when either of us thought time would run out quite like that.

Psalms 39:4-5

4 "O LORD, make me know my end and what is the measure of my days; let me know how fleeting I am!

5 Behold, you have made my days a few handbreadths, and my lifetime is as nothing before you. Surely, all mankind stands as a mere breath! Selah

Chapter 5: High school

This is one of the moments I spoke of earlier when I said I would be jumping back and forth.

You may also remember me saying relationships were hard due to those traumatic things I had been through earlier on .. even though I didn't think or realize it would impact me so much..

Or at least as much as it did..

I dated a few different girls throughout high school... I'm going to leave names out of this just for the sake of privacy and no harm to them.

One girl, I broke her heart pretty bad I would guess.. She was a real sweet girl to me..

Really controlling at the time.. but I know she cared deeply for me.. I just didn't know how to care for her on that same level and thought of her more as a friend..

It was hard to accept that someone could love me just for who I was already at 15 to 16 years old..

I already had a false sense of what love was and that it had to be earned.. and future relationships would only solidify those thoughts.

I'll call her girl number two..

Wow, man.. I really believed I loved this girl..

And we were great friends.. only I wanted more.. and she did not.. We had kinda dated briefly back in junior high school.. I was still what I thought was in love...

Looking back, I see this was the first and one of the biggest examples of this. I need to earn her love and acceptance..

Now let me be clear.. she loved me very much as a friend.. I was friend-zoned...as they say..

We passed notes all the time.. even kept a journaling book that was our friendship in a nutshell...

We had tons of laughs together.. stayed friends long into our adult lives..

But the truth is, she broke my heart. At least the heart that I still believed love was feeling, but somehow, it was a feeling that I needed to earn from others.

Then there was girl number three..

I was mesmerized by her.. captivated..

I was never sure as to why tho..

I guess cause she was different.. outgoing..

easy.. yet.. nothing ever happened with her..

She tried.. I couldn't.. I was actually still a virgin in a sense... I had never been with a girl before in that way. I had gone to second base, which is what we called it back then, but never beyond that.

I was somehow shy, if you will, in that arena despite my years before that..

So here I am.. 17 by this time.. a very confused virgin dating a girl who was nothing of the sort.

And later, I came to discover she was sleeping around behind my back cause I wouldn't perform..

Here is where God stepped in, though..

As I was dating this girl.. her foster family was actively following Jesus and attended church frequently... I started going to church with them and began attending youth group..

I grew up in a strange dynamic when it came to faith.. See my stepdad, who was a great man, by the way.. he was Catholic, and my mom had a Southern Baptist background..

My mom seldom took my sister and I to church. She did, but it was just rare.

My stepdad never went.. well, once a year at easter..

So, growing up, I heard things like.. Jesus is the reason for the season at Christmas..

But until I started to attend church at 17.. I didn't know a lot about who Jesus Christ really was..

Would I say I was saved .. like really saved at 17??

No! Not even close.. but a seed was planted!

Matthew 13:1-9

English Standard Version

The Parable of the Sower

13 That same day, Jesus went out of the house and sat beside the sea.

2 And great crowds gathered about him so that he got into a boat and sat down. And the whole crowd stood on the beach.

3 And he told them many things in parables, saying: "A sower went out to sow.

4 And as he sowed, some seeds fell along the path, and the birds came and devoured them.

5 Other seeds fell on rocky ground, where they did not have much soil, and immediately they sprang up since they had no depth of soil,

6, but when the sun rose, they were scorched. And since they had no root, they withered away.

7 Other seeds fell among thorns, and the thorns grew up and choked them.

8 Other seeds fell on good soil and produced grain, some a hundredfold, some sixty, some thirty.

9 He who has ears, let him hear

I should probably also mention my Nana and the impact she had on my life.. I loved her very much.. Sadly, it never seemed as though she could say the same thing.

I bring her up in this chapter because at the end of my junior year, as I was 17 and going to church.. she passed away.. it had been coming and wasn't a big shock. She had lung cancer about two years prior, and they got it into remission.. but at the beginning of my junior year, it came back with a vengeance and consumed her entire body..

They told my mother to take her home...keep her comfortable, and she will be gone in about six months... They were accurate in the timeline..

At her funeral, I was numb and didn't know what to feel.. I was empty.. See, my Nana picked favorites, and I wasn't one... neither was my oldest sister Patricia (Tish, we called her). My sister Jonna was the favorite out of her and I.

My older brother Mike was her favorite out of him and Tish..

Nana had no qualms, as she would most likely say, about showing it..

Our mom often told Nana not to show favorites..

Don't make it so obvious she would tell her..

Nana never paid her any mind.. (was another favorite expression she used)

Nana lived with us from the time I was probably just 7 or 8 years old until her passing..

She was something else and believed in busting you but with a switch off the tree of YOUR liking and choosing...She would say.. don't MAKE ME pick one! And she meant that!!

The things I learned from her were both great and heartbreaking.. I learned to be good and friendly around my elders.. Life can be hard.. I learned respect.. I also learned I was easy to look past, and just because someone is family.. it doesn't mean they love you..

Chapter 6: Truth

Well, truth be told.. I want to do just that.. tell you the truth and not just my side of these stories I will cover in the next few chapters..

I mean.. it is my story after all.. it would be easy to tell it as me being the victim in each and every scenario... Truth is though, I didn't know what Grace was... So I started down a path that only snowballed and got worse every time I came in contact with people... Even when I was trying to get back into church..

Let me explain.. As I was wrapping up my senior year, I was broken so badly inside but held it all in.. I played happy-go-lucky.. The life of the party if you will.. My stepdad taught me that's what men do..

Chin up.. face the world.. be a man!

I was having pre-marital sex with just about any girl I could and didn't care..

If I broke a girl's heart, I didn't care..

See at the time Frankie was still alive and my plan for graduation was to move to Morganton NC..With him and his wife Sharon.. and that is exactly what I did.. but in the meantime, I should just go ahead and have some fun right??

I wasn't in North Carolina very long before the reality of bills and not knowing anyone other than the two of them and Frankie's mom set in and I got homesick...

So my mom and dad came to get me..

I never saw Frankie again.. we barely spoke..

Five years later, Frankie was gone..

I became very angry with God!! I acted worse than Job ever did and lost way less than Job ever did..

WAIT! You are probably thinking... "Aren't you leaving out the 5 years between you coming home and Frankie's death?"...

I'm getting back to that.. right now actually..

I wasn't home for more than a couple of weeks, and one Saturday morning, I got a phone call early in the morning.. about 7...8 o'clock from my real Dad (Jack) to inform me my grandfather (his dad) had passed away.

Grandpa Tom was 85 years old and died of skin cancer and other old age complications.. nothing too serious.. It was just his time to go..

I was asked to be a pallbearer and I did so with honor, to be honest..

I loved Grandpa Tom very very much.. He always had a great story and wonderful smile to offer his grandchildren.. and you could have a banana and a glass of buttermilk.. yuk! Lol..

We really loved going to visit him!

Wasn't long after my grandfather died that I started to be very desperate for love and the meaning of life.. I would be in and out of relationships I thought were "IT".. I would also be in and out of church.. and I based my relationship with the Lord off of my happiness in worldly relationships..

I don't think, as I look back, that any of these relationships were "God-ordained"..

Each one of them had pre-marital sex in them..

Each one ended up pulling me away from church and away from God as a result..

The first one was a girl I met at the mall.. We thought.. or at least I thought we were in love (didn't know what love was at the time). We got engaged her senior year.. She was a couple years younger than me.. We spent about a year and a half together.. Her ex-boyfriend got out of prison and she dumped me.. and I spent a few months after that trying to not get killed by the guy..

Fun times I tell you..

Then I slept around for about 6 months with lots of women.. I was so lost and felt so rejected and because my life was a wreck.. I felt I couldn't turn to God.. I still didn't understand his Grace and Love.

Then I met Lisa.. the girl who would end up being the mother of my first child..

Already sounds bad huh?

Well, you are not wrong.. Lisa and I basically played games with each other the first 8 to 12 months. We knew each other.. what a clown show we were.. looking back.. it's beyond sad...

(Pause)...

During this time I had another moment. The Lord stepped in and saved my life.. and it wouldn't be the last.. He has a habit of doing that..

I was out drinking in the Uptown with an old friend of mine. Lisa and I had been fighting and that caused me to stay in the bar longer than I should have.. I made the foolish mistake of trying to drive to a friend's house where Lisa was..

I got into an accident.. hitting a few telephone poles just grazing them and then an old set of concrete steps left behind from an old building that had been taken down.. smashing in the front end of my truck and making it undrivable..

I had not realized at the moment, with all my bouncing around in the truck with a very drunk and limp body, that the arrowhead necklace I wore then had stabbed me in my neck just half an inch from my jugular vein.. I could have easily died several ways that night..

I didn't understand it at the time.. especially with the way I was living my life and sometimes I still don't. I do know now that His word is a solid promise..

Jeremiah 29:11

New King James Version

11 For I know the thoughts that I think toward you, says the Lord, thoughts of peace and not of evil, to give you a future and a hope.

That verse has so much meaning in my life now.

Back to what I was saying about Lisa..

Maybe it was that event in my life and how she cared for me that night.. cause I ran and didn't allow myself to be there when the cops showed up.. I made it to her friend's house. She cleaned me up and was a friend for the first time in our relationship.

But something happened with me.. I decided to put aside the cold calculating attitude I had toward her and give her a fair shot.. I started falling for her.. I wanted to truly be with her all of a sudden.. So I don't know.. maybe she couldn't trust that.. God knows she had been hurt a lot in the past herself..

Whatever the case might have been.. she never fully let go of her ex-boyfriend.. she ultimately ended up cheating on me with him..

As bad as that sucked.. she ended up being pregnant and not 100% sure who the father was..

I waited a long nine months to find out... trying to be there for her even though knowing it might not be my baby.. Then the day finally came and one look was all anyone needed to know he was mine.. and so.. I was torn between "thankful for the blessing" and "sadness that his mother and I would most likely never raise him together"..

It took time and a lot of rejection from Lisa before I moved on..

I guess in the grand scheme of things, I didn't waste too much time finding someone new though..

Things just started blurring together at this point in my life.. and I was like an ostrich with its head in the sand.. I didn't know how to deal with any hurt.. I just wanted band-aid after band-aid..

My Band-Aids came in the form of someone to make me feel good.. and in my mind, the only way to make me feel good was to give me sex.. because if you try to give me just your heart... well, I had learned that was fake.. Words were fake to me.. Emotional attachment was fake to me.. Sex was something I could touch.. I could feel that..

So when I met the next girl I would date.. and eventually marry.. the fact that she was a stripper was great to me.. That had sex written all over it! So again I rushed right into it.. I was so lost in the world at this point.. I had very little to do with or even think about when it came to Jesus..

This person I speak of now, her and I are not really on speaking terms.. so I won't talk a large amount about our lives together.. I will say it was an ugly marriage for the most part.. We both had our faults and made our mistakes.. We did, however, have good times and also attended church frequently.. however.. going to church doesn't make you saved and living right..

We ended up being married for 5 years.. We ended up having two beautiful children together.. I'm thankful to her for that much..

Let me pause here a moment to inject the 5-year mark that I had previously mentioned about Frankie.. During the time we were having our children.. when our oldest Morgan was just a few months old..it was Christmas time and we were at a Christmas party when I had gotten that phone call that Frankie was gone. He was hit by a drunk driver who got two years for killing him.

I had my son Bradley and she had a son before me as well named Brandon..

I loved all of them.. We saw hard times with Brandon.. He was diagnosed with non-Hodgkins lymphoma when he was 4 and went through so much.. that is actually what got us going to church.. and eventually, with lots and lots of prayers.. even his doctors said it was a miracle, that his cancer was completely in remission..

And it stayed that way..

Unfortunately, Brandon had a rough life after his mother and I divorced..

He was in and out of jail and troubled by many things.. Sadly on September 27, 2021, just 2 days after my birthday, Brandon was gunned down and murdered in his own front yard..

No one to this day has been arrested and held responsible for this heinous crime.

John 16:13

13 But when he, the Spirit of truth, comes, he will guide you into all the truth. He will not speak on his own; he will speak only what he hears, and he will tell you what is yet to come

Psalm 25:4-5

English Standard Version

4 Make me to know your ways, O Lord;

teach me your paths.

5 Lead me in your truth and teach me,

for you are the God of my salvation;

for you I wait all the day long.

This chapter was my truth as I knew it at the time.. It's how I see it still mostly through my human lens.. most of.. if not all of these events were missing His truth and His Grace.

Chapter 7: Looking Back.

I was at a point in my life after my divorce that was unfamiliar. She and I didn't do well together and that was unfortunate enough.. but I no longer saw my children daily and sometimes not at all when she decided for whatever reason she felt like..

This only led to me feeling more rejection.. more of the same old "I'm not good enough" feelings.. More of the enemy lying to me and telling me the world is better off without me.. It led to more drinking and more sex.. "fornication" if we want to call it what it really is. That went on for five years..

I dated a few girls.. even lived with one or two..

Still separated from God.. I was looking for something familiar in my life but Jesus didn't seem to be it.. at least I thought..

Then I did exactly what the Lord tells us not to do..

I looked back.. many years before this when my oldest son was just a baby and I was a big part of what we called the "Uptown bar scene".. I met and started dating a girl named Brittany.. she had a pretty big crush on me at first.. and as much as I tried not to show it.. it was fair to say I had a massive one on her as well..

But it was short-lived and we just kinda moved on.. nothing ever really happened between us.. no real sparks .. no sex.. nothing at all...

Then in 2007, our paths crossed again.. and I looked back.. We met for lunch one day and before we knew it... We were dating.. This time was different.. not in a good way.. Dating became sex. sex became living together... living together became me begging her to marry me for the next two years despite all the red flags of why that was a terrible idea.. See, Brittany was an amazing person when her depression and anxiety would ease up long enough for her to be.. Her mother had passed away when Brittany was just 16.. and she struggled with that every single day.. I tried my best to love her through it.. no easy task for a man who had so many of his own demons I was battling at the time..

Eventually the inevitable happened..

We made the obvious mistake of getting married..

And everything basically got worse and before we got to our one-year anniversary.. we were getting divorced..

This chapter is short much like it was when I was living it out..

But I honestly believe it wasn't where God was saying I was supposed to be..

Isaiah 43:18

English Standard Version

18 "Remember not the former things,

nor consider the things of old.

Philippians 3:13

English Standard Version

13 Brothers, I do not consider that I have made it my own. But one thing I do: forgetting what lies behind and straining forward to what lies ahead,

Chapter 8: Porn addict

This seems like the best place to inject my porn addiction.. This is somewhat of a hard chapter to tell.. as it is I'm sure with anyone who has suffered in this area.. It's embarrassing to admit some of these things and carries a stigma among others who claim to be perfect and never willingly admit they too have sinned and fallen short of God's glory.

1 John 1:8-10

English Standard Version

8. If we say we have no sin, we deceive ourselves, and the truth is not in us. 9. If we confess our sins, he is faithful and just to forgive us our sins and to cleanse us from all unrighteousness. 10. If we say we have not sinned, we make him a liar, and his word is not in us.

Romans 3:23

English Standard Version

23 for all have sinned and fall short of the glory of God

I'm going to start from the age of 10 when I would say what happened was molestation. I often wonder though. Did he know what he was really doing? Sounds shocking to say that right?

He was just three years older than me. 13..

Or did he? A child who killed birds with a slingshot just for the fun of it.. I don't really know what was wrong with him.. at the time it all started, he convinced me to look at pornographic material with him and masturbate next to him.. We didn't have access to the pornographic material on the Internet that one has access to nowadays.. It was the 80s.. Eventually after time, just like any other sexual sin.. he would get me to go much further.. The part that would kill me on the inside was.. Did I fight against it enough? Why didn't I tell my parents? Was something wrong with me? Why did I carry this with me for so long?

As I grew older, this grew with me.. the need in my mind to take it to the next level.. Just like his sin... my sin, too went further.. and as the times allowed for more access to more heinous porn..

I followed the enemy like a puppet on a string..

As I got older.. I brought that into relationships..

Some of the women accepted it and were very willing to participate in the filth with me.. some of them told me it was something they did before me or without me.. some of them hated it and refused to have anything to do with it...so I did it behind their backs.

Porn addiction is something that would hinder my walk with the Lord for many many years..

I fought to shake it.. thinking I had to do it to save myself.. as if my salvation was dependent on my ability to become a perfect person.. as though I had to wash myself clean and only then Jesus would accept me into His kingdom..

I ended up being addicted to porn for 37 years of my life.

1 Corinthians 6:18-20

English Standard Version

18 Flee from sexual immorality. Every other sin a person commits is outside the body, but the sexually immoral person sins against his own body. 19 Or do you not know that your body is a temple of the Holy Spirit within you, whom you have from God? You are not your own, 20 for you were bought with a price. So glorify God in your body.

Matthew 5:30

English Standard Version

30 And if your right-hand causes you to sin, cut it off and throw it away. For it is better that you lose one of your members than that your whole body go into hell.

I believe these scriptures are meant literally and figuratively.. I think the Lord is telling us it is better to lose parts of your body, people, modern-day technology, and anything else that stops you from living a sin-free life.. anything that stops you from taking up your cross and following Him.. CUT IT OFF! Cut off the Internet! Cut off the friends! Cut off whatever and whoever you need to so that you can live a life of righteousness!

I have discovered this in the last few years of my journey.. my testimony period is what I like to call it.. I am at war with the enemy.. and I no longer have the desire to do the things he can easily convince my flesh it wants to do..

Romans 7:13 Did that which is good, then, bring death to me? By no means! It was sin, producing death in me through what is good, in order that sin might be shown to be sin, and through the commandment might become sinful beyond measure. 14 For we know that the law is spiritual, but I am of the flesh, sold under sin. 15 For I do not understand my own actions. For I do not do what I want, but I do the very thing I hate. 16 Now if I do what I do not want, I agree with the law, that it is good. 17 So now it is no longer I who do it, but sin that dwells within me. 18 For I know that nothing good dwells in me, that is, in my flesh. For I have the desire to do what is right, but not the ability to carry it out. 19 For I do not do the good I want, but the evil I do not want is what I keep on doing. 20 Now if I do what I do not want, it is no longer I who do it, but sin that dwells within me.

21 So I find it to be a law that when I want to do right, evil lies close at hand. 22 For I delight in the law of God, in my inner being, 23 but I see in my members another law waging war against the law of my mind and making me captive to the law of sin that dwells in my members. 24 Wretched man that I am! Who will deliver me from this body of death? 25 Thanks be to God through Jesus Christ our Lord! So then, I myself serve the law of God with my mind, but with my flesh I serve the law of sin.

I have also discovered I am sure of the fact that I will always be a work in progress. All of this not only helps us to understand the Grace we are given.. but also to give the same Grace we have been afforded.

Philippians 1:6

English Standard Version

6. And I am sure of this, that he who began a good work in you will bring it to completion at the day of Jesus Christ.

Chapter 9: Out of the ashes

I'm including both the ESV and NKJV of these scriptures because I want you to take notice of the " instead of" and the "for" in the different translations.

A headdress instead of/ for ashes...

The oil of joy instead of/ for mourning.

The Lord is showing us here that instead of giving you what you should have.. I'm going to give you this instead..

Or even better.. I'm going to trade you..

I'll take your ashes.. and I'll give you a headdress instead.

I'll take your mourning.. and I'll give you oil of joy..

I'll take away your heaviness.. your weak spirits and give you the garment of praise so you can be called oak trees of righteousness.

Isaiah 61:3

English Standard Version

3 to grant to those who mourn in Zion—

to give them a beautiful headdress instead of ashes,

the oil of gladness instead of mourning,

the garment of praise instead of a faint spirit;

that they may be called oaks of righteousness,

the planting of the Lord, that he may be glorified.

Isaiah 61:3

New King James Version

3 To console those who mourn in Zion,

To give them beauty for ashes,

The oil of joy for mourning,

The garment of praise for the spirit of heaviness;

That they may be called trees of righteousness,

The planting of the Lord, that He may be glorified."

While I was married to Brittany, she convinced me to go back to school. Now, all though, my marriage to her failed and it being the tragedy it was..

some good things came out of those ashes..

I didn't attend a great college but I learned a lot of things and mostly, for the first time in my life, I started to build some confidence in myself due to my being on the Dean's list and actually getting the best grades of my entire schooling life..

Other things formed from that experience as well.. I met new people. New friends.. and even started going back to church.. that part.. that's the best part..

Hebrews 10:25

English Standard Version

25 not neglecting to meet together, as is the habit of some, but encouraging one another, and all the more as you see the Day drawing near.

I did not understand this verse before that.. and still not, as much as I do now,.. did I then understand it..

As I started attending this one particular church I began to fall in love with Jesus Christ again..

I started getting involved with the church.. not just attending on Sunday.. I started serving at church.. got involved with local missionary work.

I started participating in Bible study..

all the while living in sin with my current girlfriend who also attended the church.

I look back on this moment and it's a reminder to me that Christ doesn't leave us nor forsakes us.. It's us who are withdrawn from Him..

He doesn't stop loving us because we are broken.. we allow ourselves to be so broken we can't see His love.

Because of my brokenness and my desire to care more about my flesh than what The Lord was doing in my life.. I stopped attending that particular church building for a while..

That woman and I went our separate ways eventually.

I met some great people at that church and it is pretty obvious God had me there for a reason and some of those people would be there in new seasons of my life.

It wasn't much longer after this that I would meet my wife Jamie. That was at the end of 2012.

Chapter 10: Do you want to be made well?

Like most relationships, my wife and I started off well.. so it seemed to us.. but we were both pretty much living worldly lives in all truth..

We both attended church from time to time at the point we met and started dating..

But that didn't stop us from rushing into pre-marital sex and hanging out and drinking..

We didn't want to be that couple and live together in that way.. so we rushed into being married..

So after four months of knowing each other, we were married in a West Virginia courthouse. March 21st, 2013.

We were not doing great financially so we moved in with my parents and for the first three months of our marriage, we were doing wonderful..

And then who we both were.. started to surface.

See, at this point in my life.. I still was carrying around all the hurt.. all the insecurities.. all those demons.. My pride. My selfish nature.. wow.. so much inner turmoil..

Jamie was too.. she had a lot of things from her past that are her life stories to tell.. if she ever writes a book on it.. I'll let you know.. I will say..

We both had control and abandonment issues..

The abandonment was much worse for her..

I look back on this period in our marriage and think of the man at the Pool of Bethesda..

We were both two very broken people coming together.. we were spiritually blind, lame, and paralyzed.

John 5

New King James Version

A Man Healed at the Pool of Bethesda

5 After this there was a feast of the Jews, and Jesus went up to Jerusalem. 2 Now there is in Jerusalem by the Sheep Gate a pool, which is called in Hebrew, Bethesda, having five porches. 3 In these lay a great multitude of sick people, blind, lame, paralyzed, waiting for the moving of the water. 4 For an angel went down at a certain time into the pool and stirred up the water; then whoever stepped in first, after the stirring of the water, was made well of whatever disease he had. 5 Now a certain man was there who had an infirmity thirty-eight years. 6 When Jesus saw him

lying there, and knew that he already had been in that condition a long time, He said to him, "Do you want to be made well?"

7 The sick man answered Him, "Sir, I have no man to put me into the pool when the water is stirred up; but while I am coming, another steps down before me."

8 Jesus said to him, "Rise, take up your bed and walk." 9 And immediately the man was made well, took up his bed, and walked.

It would be almost 10 years later before we begin to take up our spiritual mats and walk.

As a matter of fact.. about six months into our marriage Jamie would end up going through a breast surgery that would only intensify her insecurities even though it didn't bother me one bit.. and I think that made things worse instead of better.. she saw it as me not caring instead of me not caring about the scar.. knowing her from the place I do now.. I understand.. I didn't however at that time..

Then almost a year into our marriage, she would end up injuring her back.. leaving her in need of an emergency back surgery.

That back surgery would leave her on her back for a while.. and unable to work and do housework that my wife actually likes to do most of the time..

Trying to keep her from those things was no easy task.

She eventually began to heal as best she could and had to often be reminded to turn her foot straight as she kept dealing with neuropathy and still has that at times to this day.

All that combined, with the history she was still carrying around, led her to believe lies the enemy had for her ...that she wasn't good enough for me.. that she was just a burden to me.. he told her that isn't what your husband signed up for when he married you..

Now for someone who has serious abandonment issues.. the enemy has a kill-all.. a kill-your-marriage lie.. and that is this.. leave them before they get a chance to leave you..

And Jamie bought that lie big time..

At that same time, the enemy was doing a number on me as well.. telling me.. look how ungrateful she is.. she doesn't even want to be with you.. you should just kick her out..

So I began to threaten to kick her out and started threatening to divorce her..

They were words that I didn't mean..

But that did not change their impact.. especially for a person who has abandonment issues..

Jamie had started attending college around this time for a history degree.. she started hanging out with some old friends and after a while, got reconnected with a guy she had a sexual relationship with in the past.. I wouldn't end up finding out about this for another 2 years..

Things at the time though took a very bad turn for us.. It was in the forecast already.. we were both behind cold with each other.. pushing each other away.. one night, as we were lying there in bed, she declared she was moving out..

I pleaded with her not to go.. but it slapped me in the face when she asked.. WHY NOT? You have been threatening to kick us out (her and my youngest stepson who lived with us) and been threatening to divorce me.. so why should I stay?

Unfortunately, I had no real good reason for her to stay..

Chapter 11: Broken Together

Have you ever heard the song by Casting Crowns entitled "Broken Together"?

After a few months of Jamie living about thirty-five minutes away, we decided we missed each other and should try to work it out, so I moved in with her and my stepson, Arazzio.

We almost instantly discussed that we should get back to the church..

So at first, we started attending the old church that I had mentioned just a few chapters back, which I attended right before I met Jamie. Things were okay there for a while.. but then we realized some folks we had gotten close with were no longer there.. through some conversations, we discovered they were at a new church that the youth pastor from our church left to start..

That church was literally right around the corner from our apartment.. So that became our new church as well..

Now mind you.. at no point this far had we been spending time in prayer asking God how we should move.. We were basically telling God how we were going to live.. in our actions, in our choices, and still very much in our marriage..

Jesus was a side dish if you will.. in my mind and in her mind.. We both acted as if we were the main dish.. God was not at the center of our marriage and was not number one in each of our lives.. We were going through marriage counseling at that church.. we were serving on the local missionary work team.. we were in church every Sunday morning.. We were still fighting and I still was threatening divorce.. of course after I moved in and let a couple of months go by..

Jamie has always been a person who is hard to get her attention... And I was not good at giving Grace and mercy.. I did not know what love was.

I couldn't give her something I didn't have..

In the midst of all this chaos, we did learn some things that would help us years in the future though.. We learned that we could be broken together if we really wanted to try (that wasn't yet) and we learned.. we couldn't fix each other.. that was the job of Jesus Christ and ourselves by having a one-on-one relationship with Him.. that also was not yet.. We claimed it with false pretenses.. but neither of us possessed it..

After some time, the results of our fake Christian lives were that we would end up not being in church. We would split up again and I would move out.

Jeremiah 18

New King James Version

The Potter and the Clay

18 The word which came to Jeremiah from the Lord, saying: 2 "Arise and go down to the potter's house, and there I will cause you to hear My words." 3 Then I went down to the potter's house, and there he was, making something at the wheel. 4 And the vessel that he made of clay was marred in the hand of the potter; so he made it again into another vessel, as it seemed good to the potter to make.

5 Then the word of the Lord came to me, saying: 6 "O house of Israel, can I not do with you as this potter?" says the Lord. "Look, as the clay is in the potter's hand, so are you in My hand, O house of Israel! 7 The instant I speak concerning a nation and concerning a kingdom, to pluck up, to pull down, and to destroy it, 8 if that nation against whom I have spoken turns from its evil, I will relent of the disaster that I thought to bring upon it. 9 And the instant I speak concerning a nation and concerning a kingdom, to build and to plant it, 10 if it does evil in My sight so that it does not obey My voice, then I will relent concerning the good with which I said I would benefit it.

Much like this story in Jeremiah we were broken vessels.. broken apart.. not together.. and much like the kingdoms or nations we had to choose to turn from our evil for the Lord to relent of our current disaster and be put back together.

Chapter 12: Who will pay for it?

Once I moved out of the place Jamie and I had together, it was several weeks before we spoke to each other again. It seemed like we would just move on with our lives and get a divorce.

After about three weeks, we spoke, and was I wrong?

Maybe we would be okay after all.. or would we?

One Friday evening, shortly after I got off work, I tried reaching out to see how she was doing, but there was nothing. I finally spoke to her the next day, and things were off. I feel the Lord speaks things to us even when we least expect it..

I didn't know how to act.. I got carried away with my emotions yet again and threatened divorce again.. and at this point, she definitely didn't care..

A few weeks later I found out from someone we both knew that she had moved out of that place and in with some guy..

It wasn't hard to put two and two together on what just transpired the past two months and how this guy was a part of her life longer than she let on.. and how my lack of true love, Grace, and mercy didn't help..

Truth be told.. there was no true love between us.. not true biblical love anyways..

Ephesians 5:25 Husbands, love your wives, just as Christ also loved the church and gave Himself for her, 26 that He might sanctify and cleanse her with the washing of water by the word,

I couldn't give her what I didn't have.. and neither could she give me what she didn't have.

So with that news, I began to look for who else might be out there for me.. it wasn't long before I met someone else and moved in with her..

Two can play at this game, I figured.. but what game was I really playing, and why was I okay with crushing other people in the process??

I was at a very low point in my life.. but I wasn't at the bottom yet if you can believe that or not..

Over the course of the next several months, Jamie and I would text each other for the sole reason of arguing about who would pay for our divorce..

That was all we had to disagree about.. nothing more... nothing less..

After a little while, we were making no headway on who would pay. We both somehow started to grow away from the people we were currently living with and started talking to each other with an improvement in attitude. Was the Lord stepping in?

I moved out and away from the woman I was with and temporarily in with my parents.

This was the beginning of a whole new set of problems for our marriage.

Chapter 13: Why don't you stay..?

There is a song by a country band named Sugarland.. the song entitled Stay.. at that time in our marriage would make this grown shell of a man that I was .. just lose it.. I would literally sob like a baby..

The problem was Jamie didn't want to live in my parent's house again. I'm sure due to the bad memories and due to the idea that I might abandon her and Arazzio again.. I might threaten divorce again. Could she trust me again?

The bad part was that she was living with another man. That was going to be a problem for me.

In the song Stay.. she is the other woman.. but she is being led to believe he will leave his wife for her..

So, it's not exactly the same scenario. This man is crushing two women at once, although they both don't know.

Even so.. I felt that pain this lady was singing about..

Isn't it wildly selfish how we can cause others intense pain but only have regard for our own when we are still lost in the world..

The Lord had been drawing Jamie and I to Himself, but we still had not yet truly allowed ourselves to be found.

You know, it really does seem that the Lord kept making a way for us.. no matter how many times she and I fell apart.. there was another road leading us back.. and He would gently pick us back up and put us on it..

After some time, we ended up with our own place again.. we lived in that place for about a year.. went through hard times in our marriage in that place.. We didn't attend church that whole time.. Then we moved to another place near that place.

And with the new place.. new issues arose.

I mean, we definitely had all the same arguments as before.. but in 2018, my oldest sister, Tish, passed away, and so did my Stepdad, who had raised me since I was one year old.. their passing was three months apart.

This season in my life led me to a place of drinking mostly but also smoking pot on occasion..

Also, I became very verbally abusive toward Jamie.. I would rehash the past and call her names that I thought associated her with how she made me feel back then.. I would spend weekends away from home and get plastered drunk and then text her just to argue.. just to call her names.. just to tell her I should divorce her.. and it could all be in my mind for the reason of .. I didn't like that she didn't pack my lunch on some days during the week, so she was probably cheating on me.. That went on for some time before I eventually.. you guessed it.. I moved out..

Romans 13:13

English Standard Version

13 Let us walk properly as in the daytime, not in orgies and drunkenness, not in sexual immorality and sensuality, not in quarreling and jealousy.

Ephesians 5:18

English Standard Version

18 And do not get drunk with wine, for that is debauchery, but be filled with the Spirit,

I would spend the next two years getting drunk on the weekends.. even some weekdays.. and texting my wife about what a whore she was and how she was cheating.. why? Because if I wanted to come over and see her so we could have sex and she said no for any reason.. that would be my response..

I said a couple of chapters ago that my behavior has gotten lower. I treated my wife very poorly.. was it any wonder she didn't want me around?

Chapter 14: Not my time

A time was coming for me.. what was that time, though? Time for me to die; the enemy voted..

But the Lord still had a plan for me.. so the enemy was outvoted..

Was it a time to finally be born again in my future??

Hebrews 9:27

English Standard Version

27 And just as it is appointed for man to die once, and after that comes judgment,

Ecclesiastes 3:1-8

English Standard Version

3 For everything there is a season and a time for every matter under heaven:

2 a time to be born, and a time to die;

a time to plant, and a time to pluck up what is planted;

3 a time to kill, and a time to heal;

a time to break down, and a time to build up;

4 a time to weep, and a time to laugh;

a time to mourn, and a time to dance;

5 a time to cast away stones, and a time to gather stones together;

a time to embrace, and a time to refrain from embracing;

6 a time to seek, and a time to lose;

a time to keep, and a time to cast away;

7 a time to tear, and a time to sew;

a time to keep silent, and a time to speak;

8 a time to love, and a time to hate;

a time for war, and a time for peace.

Ahh.. what a beautiful passage when consumed in its entirety.. a time for everything under the sun. Yes? I believe so...

My sister Jonna was dating a guy.. and for the respect of his family I find it not necessary to mention his name..

He and I got along just fine.. no problems amongst us at all..

I remember that even though I was still drinking and Jamie and I were still having problems.. I was thinking about getting back to church because of conversations he and I would have at times. We would get drunk and put on Christian songs we both loved..

So, one particular night, he was at the house, and we were drinking. My nephew was also there, and the two of them began to argue about something pertaining to cars, which is typical men's stuff. I was chiming in here and there with comments.. I was actually more busy eating a sandwich at the moment.. when suddenly my sister's boyfriend snuck up to me, and by the time I saw it coming, he punched me in the face.. I remember stumbling backward, and that's it until I woke up at the bottom of a flight of steps that were behind me.. my head was lying on the corner of the dresser that was at the bottom, and my head was split wide open.. ah man.. not this again..

The enemy is relentless at trying to kill people who are a threat to him..

I am told there was a whole scuffle that took place between him, my nephew, and my mom..

The police were called. And an ambulance..

He was on medication for a few different things and wasn't supposed to take one of them with alcohol for this reason.. it caused him to act out of character as he did that evening.

So when the court day came a couple of months later, I informed the judge I forgave him.. but I at least needed him to handle the doctor's bills..

Unfortunately, it seemed that he was trying to get out of that instead of just paying for it.

Sadly, a few months after that, he dropped dead of a massive heart attack..

Before he passed away, I did get back to church, and something happened between the Lord and me that had never happened before. I had a real change of mind and heart.. true repentance..

I even remember saying.. I want it to be completely different this time, Lord.. I want it to last and not walk away when it isn't going my way.. I meant that with a sincere heart!

That was in 2021.

Chapter 15: Showing the Grace of the Father.

As 2021 started for Jamie and I.. we were still living separately, and like the rest of the world.. we were faced with a new challenge.. COVID..

Seeing each other was tough.. wanting to see each other was even harder at times..

I started attending what is now our home church.

There is no doubt that God sent me to this church and had them planned out to be my church family with whom we do life. It took Jamie a while to start attending church there with me.

Almost instantly, after starting to go there, I laid down smoking and drinking.. the pornography took a while longer.. Every person who is part of this church family has shown me the love, grace, and mercy of Christ..

Each of them has prayed for Jamie and I and our marriage, and we have seen answered prayer.. we have seen walls come down..

During this time, we have also seen other blessings.. such as the opportunity to purchase my parents' house, which has been a major part of my family since I was just 19 years old..

At the time, I was working odd jobs and making wooden art and tables.. I had been a woodworker at heart for many years.. it was something my dad had a passion for, and he passed it down to me.. But I told Jamie I would have to find a full-time job to make it possible.. fearing I would be doing something I didn't have any desire to do.. I began to pray for whatever The Lord wanted me to have.. He knew my heart and must have been pleased with my obedience in the way that I put zero labels on the job and zero expectations on Him to provide a job that I wanted.. because, within two days, I saw the perfect job for me and applied.. I heard back that day.. Set up an interview for two days after that. I drove by the place the day before as something I always do to make sure I'm not late for the interview, and as I did, the Lord spoke to me and said.. they are believers in me as well.. I thought I was going crazy for a moment..

The next day, during the interview, the owner, now my boss, my friend, and my brother in Christ, revealed to me that he is also a youth pastor at his place of worship.

Praise God!!

We have seen the Lord move in big ways in our life together over the past three years..

"There are some of your graces which would never be discovered if it were not for your trials."

◇ C. Spurgeon.

Philippians 1:6

New King James Version

6 being confident of this very thing, that He who has begun a good work in you will complete it until the day of Jesus Christ

I will share this verse again with Jamie now in mind.

I understand now that she is a work in progress, too, just as I am.. I no longer want to give her the self-serving kinda Grace I once did, but I want to give her the Grace I have...

Romans 8:18

English Standard Version

18 For I consider that the sufferings of this present time are not worth comparing with the glory that is to be revealed to us.

Jamie's favorite scripture...

It is so impactful and speaks to our marriage in so many ways.. we had a lot of suffering together.

And we still will in this present life.. we are still living in a fallen world.. with a sin nature..

But none of this compares to the joy that is coming.. the glory that will be revealed to us..

And as long as we are here....

Romans 8:37

New King James Version

37 Yet in all these things, we are more than conquerors through Him who loved us.

So we can choose to let the enemy beat us down, or we can choose to love each other and put Christ at the center of our marriage..

Ephesians 6:11-18

ESV

11 Put on the whole armor of God, that you may be able to stand against the schemes of the devil.

12 For we do not wrestle against flesh and blood, but against the rulers, against the authorities, against the cosmic powers over this present darkness, against the spiritual forces of evil in the heavenly places.

13 Therefore take up the whole armor of God, that you may be able to withstand in the evil day, and having done all, to stand firm.

14 Stand, therefore, having fastened on the belt of truth and having put on the breastplate of righteousness,

15 and, as shoes for your feet, having put on the readiness given by the gospel of peace.

16 In all circumstances, take up the shield of faith, with which you can extinguish all the flaming darts of the evil one;

17 and take the helmet of salvation, and the sword of the Spirit, which is the word of God,

18 praying at all times in the Spirit, with all prayer and supplication. To that end, keep alert with all perseverance, making supplication for all the saints,

I think I know what we should choose.